Procedures for Developing Models To Predict Exceedances of Recreational Water-Quality Standards at Coastal Beaches

By Donna S. Francy and Robert A. Darner

In Cooperation with the Cuyahoga County Board of Health, Northeast Ohio Regional Sewer District, Ohio Water Development Authority, and Ohio Lake Erie Office

Techniques and Methods 6–B5

U.S. Department of the Interior
U.S. Geological Survey

U.S. Department of the Interior
DIRK KEMPTHORNE, Secretary

U.S. Geological Survey
Mark D. Myers, Director

U.S. Geological Survey, Reston, Virginia: 2006

For product and ordering information:
World Wide Web: http://www.usgs.gov/pubprod
Telephone: 1-888-ASK-USGS

For more information on the USGS--the Federal source for science about the Earth, its natural and living resources, natural hazards, and the environment:
World Wide Web: http://www.usgs.gov
Telephone: 1-888-ASK-USGS

Suggested citation:
Francy, D.S., and Darner, R.A., 2006, Procedures for developing models to predict exceedances of recreational water-quality standards at coastal beaches: U.S. Geological Survey Techniques and Methods 6–B5, 34 p.

Contents

Figures

Tables

Conversion Factors and Abbreviations

Multiply	By	To obtain
Length		
millimeter (mm)	0.03937	inch (in.)
inch (in.)	25.4	millimeter (mm)
foot (ft)	0.3048	meter (m)
Volume		
milliliter (mL)	0.06102	cubic inch (in^3)

Turbidity is reported in Nephelometric Turbidity Ratio Units (NTRU).

Concentrations of bacteria are given in colony-forming units per 100 milliliters (CFU/100 mL).

Procedures for Developing Models To Predict Exceedances of Recreational Water-Quality Standards at Coastal Beaches

By Donna S. Francy and Robert A. Darner

Abstract

State recreational water-quality standards are based on concentrations of indicator organisms, such as *Escherichia coli* (*E. coli*). Because the analytical methods for enumerating *E. coli* take at least 18–24 hours to complete, some agencies have turned to predictive modeling to obtain near-real-time estimates of recreational water quality. The USGS has been working with local agencies to develop empirical predictive models for five Lake Erie beaches in Ohio. One beach, Huntington, is used as example in this report to describe in a step-by-step fashion how data for models were collected and how models were developed and evaluated. These steps are not the only procedures that can be used to develop predictive models for beaches; rather, they are the methods used by the authors for the reported datasets.

The steps to develop predictive models are data collection; exploratory data analysis; model development, selection, and diagnosis; determination of model output values; and model validation and refinement. For Huntington, the predictive model was based on data collected during the recreational seasons of 2000–2004. The explanatory variables were wave height, weighted rainfall in the past 48 hours, and $\log_{10}$ turbidity; the model explained 38 percent of the variability in *E. coli* concentrations. Two outputs from the model were calculated: (1) the predicted *E. coli* concentration and (2) the probability that the *E. coli* single-sample maximum bathing-water standard of 235 colony-forming units per 100 milliliters (CFU/100 mL) will be exceeded. A threshold probability of 29 percent was established for the Huntington 2000–2004 model. The threshold probability is the probability associated with too great a risk to allow swimming and is established by examining historical data. The model was validated in 2005 and yielded more correct responses and better predicted exceedance of the bathing-water standard than did the current method for assessing recreational water quality (using the previous day's *E. coli* concentration).

The procedures described in this report can be used to develop and test predictive models at other beaches. Predictive modeling is a dynamic process meant to augment existing beach-monitoring programs, not to replace them. Models should be continuously validated and refined to improve predictions and better protect public health. If validation tests are successful, a beach manager may decide to develop an Internet-based system that provides model predictions to the beach-going public. This type of system, called "nowcasting," was implemented at Huntington on May 30, 2006.

Introduction

As the result of the Beaches Environmental Assessment and Coastal Health (BEACH) Act of 2000, states have adopted U.S. Environmental Protection Agency criteria into state recreational water-quality standards (U.S. Environmental Protection Agency, 1986, 1998, 1999). These include concentrations of bacterial indicators—*Escherichia coli* (*E. coli*) or enterococci for freshwaters and enterococci for marine waters. The analytical methods for these organisms, however, take at least 18–24 hours to complete. Recreational water-quality conditions may change during this time, leading to erroneous assessments of public-health risk. As a result, some agencies have turned to predictive modeling to obtain near-real-time estimates of recreational water quality. In situations where nonpoint or unidentified sources dominate, empirical modeling is most appropriate. Empirical models, developed through statistical techniques such as multiple linear regression (MLR), use easily measured environmental and water-quality variables to estimate bacterial-indicator concentrations or the probability of exceeding target concentrations.

Researchers have worked to study the use of environmental and water-quality variables and to develop empirical models for assessments of recreational water quality for coastal waters. Ackerman and Weisberg (2003) confirmed the use of a rainfall advisory system in southern California and found that rainfall amounts greater than 6 mm consistently led to beach water-quality degradation. At recreational estuaries in Australia, rainfall alone accounted for 15–66 percent of the variability in bacterial-indicator densities (Hose and others, 2005). In a study in Indiana, several variables measured with onsite sensors were used to develop equations that accounted for 60–90 percent of the variability in *E. coli* concentrations at the outlets of two ditches draining into Lake Michigan (Olyphant and others, 2003). Explanatory variables used in the Indiana study were rainfall, stream discharge, soil temperature, depth to the water table, and nitrate and ammonia concentrations. In this same area, a model was developed to predict recreational water quality at five proximate beaches affected by fecal contamination from the same ditch (Whitman, 2005). Similarly, sensors to measure hydrometeorological variables were deployed at two beaches in northern Illinois and designated as "SwimCast" methodology by investigators. The SwimCast models were more successful than the previous day's bacteria concentrations at predicting whether the beach was safe for swimming (Olyphant and Pfister, 2005). Beach-specific models were developed for Ohio Lake Erie beaches using MLR techniques and 1 or 2 years of data (Francy and Darner, 2002; Francy and others, 2003). The explanatory variables included wave height, number of birds on the beach, lake-current direction, rainfall, turbidity, and wind direction and speed.

The U.S. Geological Survey (USGS), in cooperation with the Cuyahoga County Board of Health, the Northeast Ohio Regional Sewer District, the Ohio Water Development Authority, and the Ohio Lake Erie Office, has been working to develop predictive models for five Lake Erie beaches in Ohio: Lakeview (Lorain, Ohio), Huntington Reservation (Bay Village, Ohio), Edgewater Park and Villa Angela (Cleveland, Ohio), and Lakeshore Park (Ashtabula, Ohio) (fig. 1). At Huntington Reservation ("Huntington"), investigations are further along than at other beaches. In this report, we describe how data for the models were collected and how the models were developed and evaluated, using Huntington as an example. Procedures are discussed in a step-by-step fashion so that they can be used by beach managers, scientists, and others in other coastal areas to develop predictive models for local beaches. The steps in this report are by no means the only procedures that can be used to develop predictive models for beaches; rather, they are the methods used by the authors for the reported datasets and were included to make the procedures as simple and easy to follow as possible.

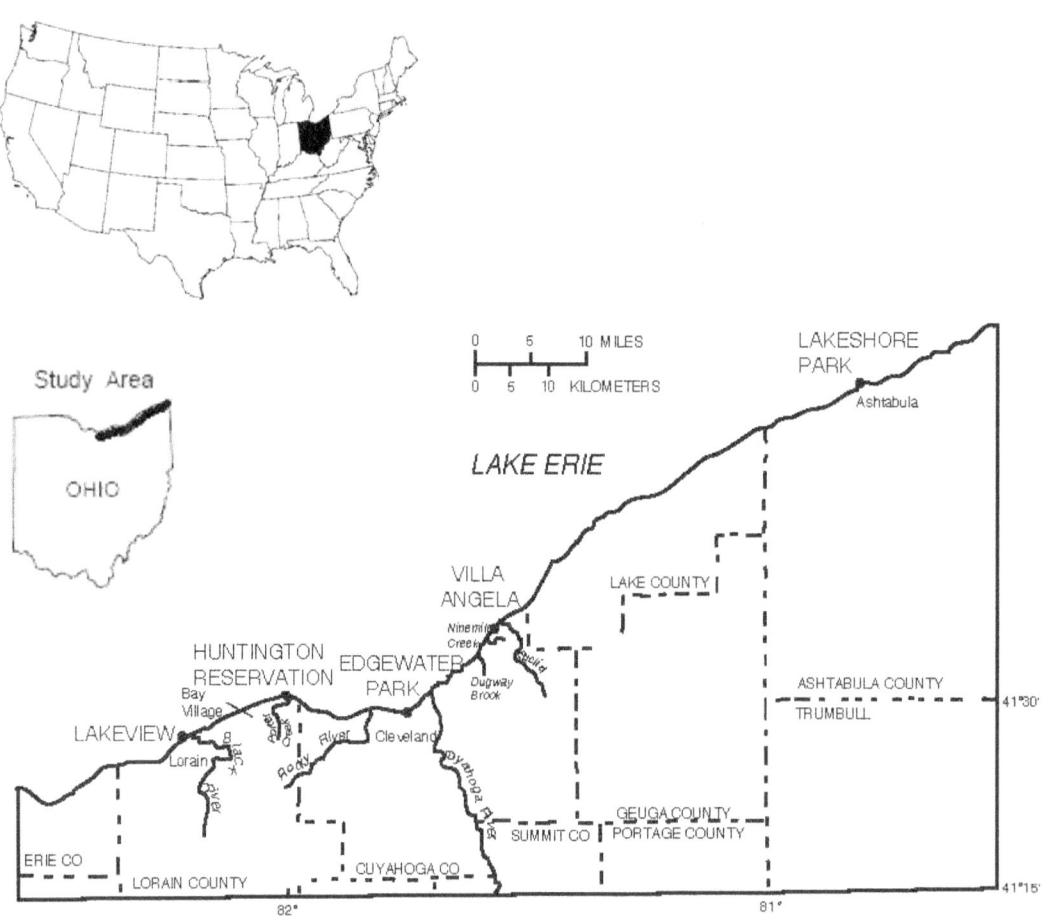

Figure 1.   Locations of five Lake Erie beaches used to test development of predictive models.

Procedures for Developing Predictive Models

Procedures for developing predictive models involve data collection, exploratory data analysis, model development, model diagnostics and selection, and model output and validation. (Use the "go to " links to jump to illustrative examples.)

Data Collection

The most important step in any monitoring program is to collect a high-quality data set. At least two seasons of data should be collected for predictive-model development, and data from a third season should be used for model validation. Weather conditions, recreational use, and lake levels can differ from year to year, so collecting only one season of data may lead to development of predictive models that are calibrated on a small subset of possible environmental conditions. Ideally, data for predictive-model development should be collected 7 days a week; however, economic considerations often preclude such an intensive data-collection effort. At the very least, four consecutive days of data are needed each week. In order to present water-quality information to the public before they head to the beach, samples are preferably collected in the morning.

Collect samples for *E. coli* and turbidity in the area used for swimming; at some beaches more than one sample may be needed to represent water-quality conditions at the beach. Collect samples in knee- to waist-deep water, consistently from the same depth each day. Be sure to leave about 1 to 2 in. of headspace in each bottle to allow adequate mixing. Place samples on ice immediately after collection and analyze them within 6 hours. Detailed information about establishing beach sites, designing a monitoring program, and collecting samples is available in U.S. Environmental Protection Agency (2002a, chap. 4 and Appendix J).

Go to example 1

Samples should be analyzed for *E. coli* and enterococci by use of a USEPA-recommended method (U.S. Environmental Protection Agency, 2002a, p. 4–17 to 4–19). Membrane filtration (MF) and most probable number (MPN) are the two types of methods that are currently used for enumerating bacterial indicators in ambient waters (U.S. Environmental Protection Agency, 2002a, p. 4-17).

If more than one sample is collected at each beach (multiple-point samples), the sample can be analyzed separately or composited. If multiple-point samples are analyzed separately, calculate the daily average indicator concentration. (Average concentrations are used instead of median concentrations so as not to downweight the influence of extreme values.) To form a composite, shake each point sample to ensure homogeneity of the sample. Immediately after shaking, combine 100-mL aliquots from each point sample into a sterile bottle to form the composite sample. In a recent study of Lake Erie beaches, *E. coli* concentrations from averaged multiple-point samples and from composite samples were not significantly different and yielded similar measures of recreational water quality (Erin E. Bertke, U.S. Geological Survey, written commun., 2006).

Ideally, turbidity should be measured onsite by use of a field turbidimeter or in situ by use of a water-quality meter. If the sample is transported to a laboratory for turbidity analysis, be sure to keep the sample on ice at all times. Because turbidity instruments of different designs may not yield equivalent results, use the same instrument throughout the project (Anderson, 2005).

Go to example 2

Because models are only as good as the data used to develop them, strict quality-assurance and quality-control (QA/QC) practices are essential. Distribute field and laboratory protocols to all personnel to ensure that procedures are followed correctly and consistently. Do onsite QA/QC checks of procedures performed by field and laboratory personnel throughout the recreational season. Procedures for QA/QC laboratory practices are described in Francy and others (2005).

Quality-control samples are collected to measure sampling and analytical variability or contamination potential. At least 10 percent of *E. coli* samples should be QC samples including split replicates, field blanks, and positive-control reference cultures. Split replicates consist of two bottles collected by the same person at the sampling point, each bottle being analyzed twice. Field blanks measure contamination potential during sample collection and handling. To collect a field blank, pour 200–500 mL of sterile buffer into the bottle under actual field conditions. Positive-control reference cultures are pure cultures of *E. coli* obtained from a commercial supplier or prepared in house. Carefully monitor results from QC samples, and retest and (or) take corrective measures when needed. For turbidity, measure duplicate aliquots from the same bottle, and repeat measurements that do not agree within established control limits. Send turbidity reference standards periodically to field or laboratory personnel analyzing the samples so that instruments and techniques can be checked for accuracy.

Go to example 3

Obtain information from recently conducted sanitary surveys, talk to local water-resource managers, and (or) visit the beach several times to compile a list of possible variables that may be affecting bacterial-indicator concentrations at the beach. These are the variables that are used to develop predictive models.

- Obtain wave-height data of some kind, because wave height is an important variable at most Great Lakes beaches (Francy and others, 2003: Olyphant and Pfister, 2005). Wave heights can be estimated visually and placed into categories, measured by use of a graduated stick, or measured by use of in situ equipment; more precise measurements may lead to better predictive models.

- Count or carefully estimate the number of birds on the beach from a remote location so as not to disturb them.

- Obtain weather (rainfall and wind direction) and lake-level data from the nearest reliable source; in most cases, weather data are available from a nearby airport. If local weather data are not available, consider installing a rain gage or weather station near the beach or within the watershed.

- Obtain any other reliable data that may be available, such as streamflow from nearby tributaries (U.S. Geological Survey, 2006) and effluent discharge information from wastewater-treatment plants or combined-sewer overflows.

Go to example 4

Exploratory Data Analysis

For each beach, daily data are compiled into spreadsheets and reviewed at least weekly throughout the recreational season so that any errors can be quickly addressed. Equations and spreadsheet entries are checked by a second person. Because of the wide range of expected values, bacterial concentrations are generally $\log_{10}$ transformed before exploratory data analysis.

A good way to start is to examine summary statistics by year and for multiple years of data combined, including the median, minimum, and maximum bacterial-indicator concentrations and the number of the days the standard was exceeded. This will provide general water-quality information and may help to explain between-year differences in important explanatory variables.

Go to example 5

After examining the yearly summary statistics, construct graphs of the bacterial-indicator concentrations and possible explanatory variables. Scatterplots are used to examine the relation between a continuous variable, such as rainfall, and bacterial-indicator concentrations and to ensure that the relation is linear. Plot each explanatory variable on the x-axis and average concentrations of *E. coli* or enterococci on the y-axis. Plots may indicate relations that are nonlinear. In those cases, options are to find a linearizing transformation, con-

sider expressing the variable in categories, or omit the variable for inclusion in the linear model. Consider using boxplots to understand the distribution of indicator concentrations as a function of variables that are not continuous but rather are grouped by categories, such as wave height and wind direction. Analyze plots by year and for all years combined.

Go to example 6

After graphs have been constructed and analyzed, statistics are calculated to quantify the strength of the associations between bacterial-indicator concentrations and possible explanatory variables and to understand the relations among explanatory variables. A significance level of $\alpha < 0.05$ is a default value generally used in traditional statistics, but there is no reason why other values should not be used (Helsel and Hirsch, 2002, chap. 4, p. 106–107). The significance level is the risk deemed acceptable by the decision maker of rejecting the null hypothesis when it is in fact true.

Pearson's r may be used to determine the linear association between bacterial-indicator concentrations and continuous variables (Helsel and Hirsch, 2002, chap. 8, p. 209); the null hypothesis, in this case, is that the correlation coefficient is zero. The Pearson's r correlation coefficient helps identify which variables are possible important predictors of bacterial-indicator concentrations. Pearson's r correlation coefficients may also be used to determine the relations among explanatory variables; explanatory variables that are strongly related are "collinear" and may reduce the strength of a model.

Analysis of variance (ANOVA) may be used to determine the relations between categorical variables and bacterial-indicator concentrations; indicator data are placed into groups on the basis of variable categories. If bacterial-indicator data are not normally distributed, they are combined and ranked from lowest to highest and an ANOVA is computed on the ranks; this is a nonparametric ANOVA (Helsel and Hirsch, 2002, chap. 7, p. 157–163). If the ANOVA indicates differences between groups, the Tukey-Kramer multiple comparison test can be used to determine which groups differ from each other (Helsel and Hirsch, 2002, chap. 7, p. 195–200). In ANOVA, the null hypothesis is that an explanatory variable is not related to the bacterial-indicator concentration. Results from ANOVA show which variables are important and serve as guidelines in grouping categorical data for model development.

Go to example 7

Graphs, correlations, and results from ANOVA and Tukey's test are then examined together to determine the following:

- Which explanatory variables had strong associations with bacterial-indicator concentrations? These include continuous or categorical variables with significant correlations to indicator concentrations or significant differences in bacterial-indicator concentrations among groups, respectively. These are the variables that should be given a higher priority for inclusion in the model.

- Was the relation consistent from year to year? These include variables with significant statistics from year to year and similar graphical relations. A consistent relation indicates that the model likely will do better in predicting bacterial-indicator concentrations in subsequent years than models employing explanatory variables that do not show consistent relations.

- Was the relation influenced by one or two extreme values? If so, were these extreme values valid measurements? After examining graphs for outliers, determine whether an outlier resulted from sampling and (or) analytical errors. If an outlier value was erroneous, omit it from the model. If the outlier appears to be a valid measurement, it should remain in the dataset; data collected in subsequent years may verify the validity of the outlier.

- Is the relation between the explanatory variable and *E. coli* linear? A linear relation can be identified by examining graphs of continuous data. If the data do not appear to lie along the straight line and some other pattern is evident, consider doing a data transformation of the explanatory variable that results in a linear relation, or omit the variable altogether (Helsel and Hirsch, 2002, chap. 9, p. 228–229).

- Were two explanatory variables strongly correlated to each other (collinear)? These include continuous variables with significant relations to each other, such as two rainfall variables. It can also be two different variables that are correlated, such as lake level and day of the year. Consider using only one of the strongly correlated variables in the model.

Model Development

Explanatory variables that show significant relations to bacterial indicator concentrations are used to produce a list of possible MLR models. A means of doing this is the Mallows' Cp test (Mallows, 1973), by which the MLR models are ordered so that R^2 is maximized and Mallows' Cp statistic is minimized. The R^2 of each model (coefficient of determination) is the fraction of the variation in *E. coli* concentrations that can be explained by the given combination of explanatory variables. The Cp statistic is a measure of the error in a model with a subset of explanatory variables relative to the error in a model that incorporates all potential explanatory variables. The steps in model development using the Mallows' Cp test are as follows:

1. Include all variables with significant relations to indicator concentrations as potential variables in the model.

2. Produce an ordered list of models based on R^2 and Mallows' Cp values. An example of commands used in a statistical package to determine the best 50 models is shown in Appendix 1, Example 1.1.

Models are then selected for further examination on the basis of the Mallows' Cp ranking and a subjective component by considering the following:

- Does the model include explanatory variables that are strongly related to each other (collinear)? Collinearity may destabilize the MLR equation. If so, consider eliminating these types of variable combinations from the model, especially if no additional information is gained from including related variables.

- Does the model include variables that are difficult or not practical to measure? If so, include those variables only if they significantly improve the model.

- Does the model include an explanatory variable that was shown to be important in exploratory data analysis? Among similar Mallows' Cp statistics, select the model that includes the variable or variables you feel are important at a particular beach.

Go to example 8

Model Diagnostics and Selection

Model statistics are examined and diagnostic tests are done to identify the model(s) for each beach that provide the best linear, unbiased estimator of bacterial-indicator concentrations (Helsel and Hirsch, 2002, chap. 9, p. 228–237). These include determination of parameter estimates, Cook's D values, partial residual plots, and residual plots. An example of commands used in a statistical package to determine these model parameters is shown in Appendix 1, Example 1.2; an example of the output is shown in Example 1.3. Performing well on model diagnostic tests and having a set of explanatory variables that seems reasonable and are relatively easy to collect are the criteria for choosing the "best" model for each beach. To determine the best model, consider the following:

- Are the parameter estimates reasonable in magnitude and significant for each explanatory variable? Sign and magnitude of the parameter estimate should be in keeping with the expected effect of the explanatory variable on the predicted variable. The t-values and p-values provide information on the significance of each parameter estimate and indicate whether or not the parameter is different from zero (Helsel and Hirsch, 2002, chap. 9, p. 237–238). Variables with parameter estimates that are not significant should be considered for elimination.

- Do any observations have high influence on the regression? Cook's D is used as a measure of influence and leverage. The critical Cook's D is calculated from the number of explanatory variables and observations. An MLR with several explanatory variables and more than 30 observations would have a critical Cook's D value in the range of 1.6 to 2.0 (Helsel and Hirsch, 2002, chap. 9, p. 248). A value higher than the critical range indicates that the observation has high influence on the regression. If this occurs, examine the value for possible errors or special conditions that may preclude its inclusion in the dataset. For example, if the observation forces a best-fit line away from a large portion of the data, consider omitting it from the dataset; this can be seen through a plot of measured versus predicted indicator concentrations.

- Do the partial residual plots indicate each variable is influencing the regression? In a partial residual plot, the bacterial-indicator concentration is regressed against all explanatory variables except for one, and the residuals are plotted against the omitted explanatory variable. These plots show how much influence the omitted variable has on the regression by eliminating the effects from other variables. If the partial plot for any variable does not show an expected pattern (linear for continuous variables, increasing or decreasing for categorical variables), consider a transformation for the explanatory variable or using an alternative model.

- Is the relation linear? Plot measured versus predicted indicator concentrations to ensure the relation is linear. Examine any outliers from the graph for data errors or commonalties. For example, if the outliers occurred on days when wave height was elevated, and wave height is not an explanatory variable in the model, consider using a model that includes wave height.

- Are the residuals evenly distributed around the zero-residual line over the range of observations? Regression residuals are plotted against predicted bacterial-indicator concentrations to determine whether residuals are similar in range and evenly distributed above and below the zero line over the entire range of observations. If they were not, consider transforming a variable, adding an additional variable, or selecting an alternative model. Examine the relation between regression residuals and date to look for autocorrelation; tests and remedies for autocorrelation are described in Montgomery and Peck (1982).

Go to example 9

Model Output and Validation

Two types of output may be produced by the MLR models. The first and obvious output is the predicted bacterial-indicator concentration. Because prediction intervals have been shown to be fairly wide in earlier studies (Francy and Darner, 1998; Francy and others, 2003), a second output variable may be used in the hope of getting a more accurate prediction—the probability of exceeding an appropriate target value. For the USGS studies in Ohio to date, the target has been exceedance of the single-sample maximum bathing-water standard. The probability that the predicted value is greater than 235 CFU/100 mL is computed as the probability of Student's t being greater than x, with the degrees of freedom equaling the number of observations used in the regression minus the number of regression coefficients in the regression equation.

$$x = (\log(235) - \hat{y} / \text{sep}$$

where $\hat{y}$ is the regression estimate of the $\log_{10} E.\ coli$ and sep is the standard error of prediction of y.

For each selected model, a probability associated with too great a risk to allow swimming is determined retrospectively—this is called the threshold probability. Threshold probabilities are determined by taking the dataset used to develop the model and finding the probability that is a reasonable balance between achieving a high number of correct responses and a low number of false negative responses. Computed probabilities that are less than the threshold indicate that bacterial water quality is most likely acceptable for swimming. Computed probabilities equal to or greater than the threshold probability indicate that the water quality is most likely not acceptable and that a water-quality advisory may be needed. Model specificities and sensitivities for the threshold-probability technique are reported and compared to specificities and sensitivities associated with the current method used to assess recreational water quality. The sensitivity is the proportion of actual exceedances (concentrations > 235 CFU/100 mL) that are predicted correctly (by the model or the current method) as being above the standard. The specificity is the proportion of nonexceedances that are correctly predicted as being below the standard.

Go to example 10

Models perform fairly well when predicting responses to data used to develop them. A better test of a model is to predict responses for an independent period. For model validation, data are collected during an independent year (a year whose data were not used for model development) to compare the model's performance with the current method for assessing recreational water-quality. After validation tests, the additional year of data can be added to the model-development process, and a new model with another year of data is developed for use in subsequent years.

Go to example 11

The Future of Predictive Modeling

The procedures described in this report can be used to develop predictive models at local beaches; all that is needed is an existing monitoring program, a basic knowledge of statistics, and computer software. Equipment costs for data collection are minimal, because most of the data required for predictive models are available from other agencies or are easily measured by field technicians at the beach. As a model proves to be a useful tool at a particular beach, beach managers may decide to invest in more expensive equipment to measure environmental conditions in real time. Also, if validation tests are successful, beach managers may also decide to develop an Internet-based system that provides model predictions to the beachgoing public. An Internet-based system enables a beach manager to provide reliable estimates of recreational water quality on weekends. Currently, weekend estimates are not commonly available because of the time and cost of laboratory analysis.

Predictive modeling is a dynamic process meant to augment existing beach-monitoring programs, not to replace them. Models should be continuously validated and refined to improve predictions and better protect public health.

Go to example 12

Summary

State recreational water-quality standards are based on concentrations of indicator organisms, such as *E. coli*. Because the analytical methods for enumerating these organisms take at least 18–24 hours to complete, some agencies have turned to predictive modeling to obtain near-real-time estimates of recreational water quality. Empirical predictive models, developed through statistical techniques such as multiple linear regression, use easily measured environmental and water-quality variables to estimate bacterial-indicator concentrations or the probability of exceeding target concentrations.

The USGS has been working with local agencies to develop empirical predictive models for five Lake Erie beaches in Ohio. At Huntington Reservation, Bay Village, Ohio, investigations are further along than at other beaches; six years of data have been collected and a model has been validated during an independent year. In this report, Huntington is used as example to describe how data for models were collected and how models were developed and evaluated. Procedures are discussed in a step-by-step fashion so that they can be used by beach managers, scientists, and others in other coastal areas to develop predictive models for local beaches. The steps in this report are by no means the only procedures that can be used to develop predictive models for beaches; rather, they are the methods used by the authors for the reported datasets.

The steps to develop predictive models include data collection, exploratory data analysis; model development, selection, and diagnosis; determination of model output values; and model validation and refinement. At Huntington, data were collected or compiled during the recreational seasons of 2000–2005 to determine *E. coli* concentrations, turbidity, bird counts, water temperature, categorical wave heights, lake levels, rainfall amounts, and wind directions. A predictive model was developed for the 2000–2004 data; this model was validated in 2005, and a new model was developed from 2000–2005 data.

During exploratory data analysis at Huntington, correlations between *E. coli* concentrations and explanatory variables showed that the strength and significance of correaltions can differ from year to year for some variables, whereas other variables were consistently and significantly related to *E. coli*. At Huntington, R_{d-1}, turbidity, and $\log_{10}$ turbidity were positively and significantly related to *E. coli* for all years tested. Combining two days of rainfall data (Rw48) improved the correlations to *E. coli* over single-day rainfall variables. Categorical data were examined by use of boxplots and analysis of variance. *E. coli* concentrations increased with increasing wave height but were not significantly related to wind direction.

A list of possible models, along with their Mallows' Cp statistic and R^2 values were developed for the Huntington 2000–2004 data. The best model contained the variables wave height, Rw48, and $\log_{10}$ turbidity and explained 38 percent of the variability of *E. coli* concentrations (Huntington 2000–2004 model). The Huntington 2000–2004 model passed regression diagnostic and hypothesis tests. Two outputs from the model were calculated: (1) the predicted *E. coli* concentration and (2) the probability that the single-sample maximum bathing-water standard of 235 CFU/100 mL *E. coli* will be exceeded. A threshold probability of 29 percent was established for Huntington 2000–2004 model. The model was validated in 2005 and yielded more correct responses and better predicted exceedance of the bathing-water standard than the current method for assessing recreational water quality (use of the previous day's *E. coli* concentration). In fact, the current method failed to accurately predict any of the eight exceedances, whereas the model accurately predicted four of them. A new model based on 2000–2005 data was developed that explained 42 percent of the variability of *E. coli* concentrations and included the same three variables plus day of the year (Huntington 2000–2005 model). Predictions based on the Huntington 2000–2005 model and the threshold probability have been presented to the public through an Internet-based "nowcasting" system since May 30, 2006; the model will continue to be validated and refined.

Predictive modeling is a dynamic process meant to augment existing beach-monitoring programs, not to replace them. Models should be continuously validated and refined to improve predictions and better protect public health. The procedures described in this report can be used to develop predictive models at other local beaches.

Acknowledgments

Individuals from many agencies helped to ensure the successful completion of this study. The authors thank Lester Stumpe, Mark Citriglia, and Eva Hatvani of the Northeast Ohio Regional Sewer District; Jill Lis of the Cuyahoga County Board of Health; Jack Kurowski of the Lorain City Health Department; Brenda Stephens of the Ashtabula Township Park Commission; and students Lena Kavaliauskas, Timothy Roberts, and Paula Carver. Special thanks go to Greg Koltun and Tammy Zimmerman for their statistical expertise and helpful reviews of this report.

References Cited

[Online references active at the time of publication are linked to those Web sites in this document.]

Ackerman, D., and Weisberg, S.B., 2003, Relationship between rainfall and beach bacterial concentrations on Santa Monica Bay beaches: Journal of Water and Health, v. 1, no. 2, p. 85–89.

Anderson, C.W., 2005, Turbidity: U.S. Geological Survey Techniques of Water-Resources Investigations, book 9, chap. A6., section 6.7, accessed June 2006 at *http://pubs. water.usgs.gov/twri9A6/*

Finney, R.L., and Thomas, G.B., 1989, Calculus: Reading, Mass., Addison-Wesley Publishing Company, chaps. 1.5 and 11.

Francy, D.S., and Darner, R.A., 1998, Factors affecting *Escherichia coli* concentrations at Lake Erie public bathing beaches: U.S. Geological Survey Water-Resources Investigations Report 98–4241, 41 p.

Francy, D.S., and Darner, R.A., 2002, Forecasting bacteria levels at bathing beaches in Ohio: U.S. Geological Survey Fact Sheet FS–132–02, 4 p.

Francy, D.S., Gifford, A.M., and Darner, R.A., 2003, *Escherichia coli* at Ohio bathing beaches—Distribution, sources, wastewater indicators, and predictive modeling: U.S. Geological Survey Water-Resources Investigations Report 02–4285, 120 p.

Francy, D.S., Bushon, R.N., Brady, A.M.G., Kephart, C.M., and Stoeckel, D.M., 2005, Quality assurance/quality control manual for the Ohio Water Microbiology Laboratory, accessed at *http://oh.water.usgs.gov/micro/lab.html#qcm*

Helsel, D.R., and Hirsch, R.M., 2002, Statistical methods in water resources: U.S. Geological Survey Techniques of Water-Resource Investigations, book 4, chap. A3, accessed March 2006 at *http://pubs.er.usgs.gov/pubs/twri/twri04A3*

Hose, G.C., Gordon, G., McCullough, F.E., Pulver, N., and Murray, B.R., 2005, Spatial and rainfall related patterns of bacterial contamination in Sydney Harbour estuary: Journal of Water and Health, v. 3, no. 5, p. 349–358.

Mallows, C.L., 1973, Some comments on Cp: Technometrics, v. 15, p. 661–675.

Montgomery, D.C., and Peck, E.A., 1982, Introduction to linear regression analysis: New York, John Wiley & Sons, p. 347–360.

Myers, D.N., and Wilde, F.D., eds., 2003, Biological indicators (3d ed.): U.S. Geological Survey Techniques of Water-Resources Investigations, book 9, chap. A7, accessed March 2006 at *http://pubs.water.usgs.gov/twri9A7/*

National Oceanic and Atmospheric Administration, 2005a, International Great Lakes Datum: Center for Operational Oceanographic Products and Services, accessed October 2005 at *http://www.co-ops.nos.noaa.gov/*

National Oceanic and Atmospheric Administration, 2005b, National Virtual Data System—National Climatic Data Center: Asheville, N.C., accessed October 2004 and 2005 at nndc.noaa.gov/

Ohio Environmental Protection Agency, 2003, Water use definitions and statewide criteria: Ohio Administrative Code, chap. 3745-1-07, p. 7 and 26, accessed April 2006 at *http://www.epa.state.oh.us/dsw/rules/3745-1.html*

Olyphant, G.A., and Pfister, M., 2005, SwimCast—Its physical and statistical basis, *in* Proceedings of the Joint Conference—Lake Michigan State of the Lake and the Great Lakes Beach Association, Green Bay, Wis., November 2–3, 2005: p. 56, accessed at *http://www.great-lakes.net/glba/2005conference.html*

Olyphant, G.A., Thomas, J., Whitman, R.L., and Harper, D., 2003, Characterization and statistical modeling of bacterial (*Escherichia coli*) outflows from watersheds that discharge into southern Lake Michigan: Environmental Monitoring and Assessment, v. 81, p. 289–300.

SAS Institute, Inc., 1990, SAS/STAT user's guide, version 6 (4th ed.): Cary, N.C. [multiple volumes].

U.S. Environmental Protection Agency, 1986, Ambient water quality criteria for bacteria—1986: Washington, D.C., Office of Research and Development, EPA–440/5–84–002, 18 p.

U.S. Environmental Protection Agency, 1998, Bacterial water quality standards for recreational waters—Status report: Washington, D.C., Office of Water, EPA–823–R–98–003.

U.S. Environmental Protection Agency, 1999, EPA action plan for beaches and recreational waters: Washington, D.C., Office of Water, EPA–600–R–98–079.

U.S. Environmental Protection Agency, 2000, Improved enumeration methods for the recreational water quality indicators—Enterococci and *Escherichia coli*: Washington, D.C., Office of Science and Technology, EPA/821/R–97–004, 49 p.

U.S. Environmental Protection Agency, 2002a, National beach guidance and required performance criteria for grants: Washington, D.C., EPA 823–B–02–004 [variously paginated].

U.S. Environmental Protection Agency, 2002b, Method 1603—*Escherichia coli* in water by membrane filtration using modified membrane-thermotolerant *Escherichia coli* agar: Washington, D.C., Office of Water, EPA 821–R–02–23, 9 p.

U.S. Environmental Protection Agency, 2004, Water quality standards for coastal and Great Lakes recreational waters—Final rule: Federal Register 40 CFR Part 131, v. 69, no. 220, p. 67,217–67,226.

U.S. Geological Survey, 2006, Real-time water data for the nation, accessed June 2006 at *http://waterdata.usgs.gov/nwis/rt*

Whitman, Richard, 2005, Project S.A.F.E., accessed August 2006 at *http://www.glsc.usgs.gov/projectSAFE.php*

Examples From Beach Studies at Huntington Reservation, Bay Village, Ohio

Example 1

Data were collected at Huntington, Bay Village, Ohio, during the recreational seasons (May through September) of 2000–2005. Data from 2000–2004 were used to develop a predictive model, and data from 2005 were used to validate the model. Samples were collected by the Cuyahoga County Board of Health (CCBH), Monday through Thursday mornings at two sampling points in the swimming area in thigh-deep water using a grab-sampling technique (Myers and Wilde, 2003). Samples were collected in 500-mL autoclaved polypropylene bottles, with 1 to 2 in. of headspace in each bottle for proper mixing. The bottles were placed on ice within 10 minutes of sample collection.

Central sampling location, Huntington Reservation, Bay Village, Ohio. (Photo by Donna Francy, U.S. Geological Survey.)

Samples are collected in a 500-mL autoclaved polypropylene bottle using a grab-sampling technique. (Photo by Donna Francy, U.S. Geological Survey.)

Back to page 3

Example 2

At Huntington, water samples were transported to the Cuyahoga County Sanitary Engineers laboratory and analyzed for *E. coli* and turbidity within 3 hours of collection. In Ohio, *E. coli* concentrations are used to monitor recreational water quality. Each point sample was analyzed for concentrations of *E. coli* by use of the mTEC (U.S. Environmental Protection Agency, 2000) or modified mTEC (U.S. Environmental Protection Agency, 2002b) membrane-filtration methods. A daily *E. coli* concentration was calculated by averaging results from two sampling points. An aliquot of water was removed from the bottle to measure turbidity in the laboratory by use of a turbidimeter.

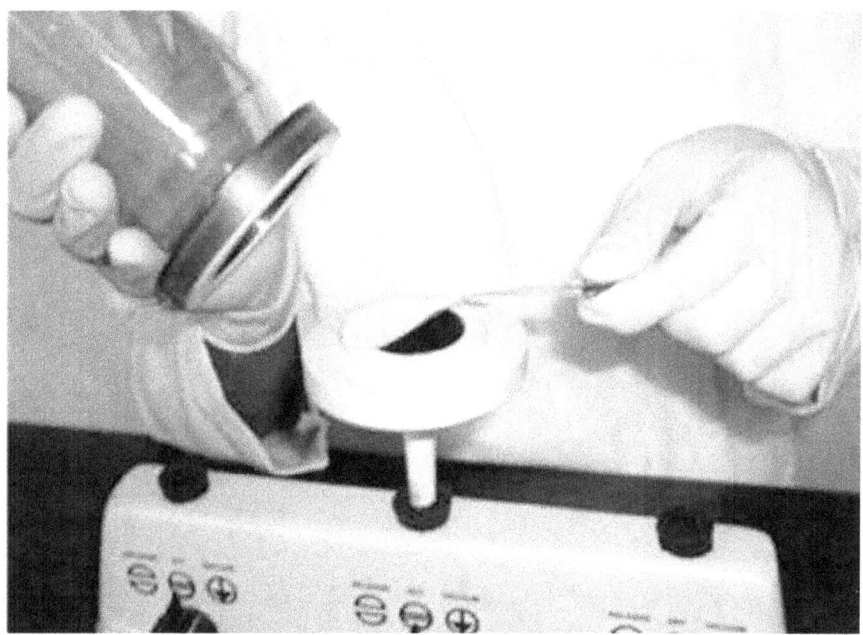

Filter being applied to suction device during membrane-filtration procedure. (Photo by Donna Myers, U.S. Geological Survey.)

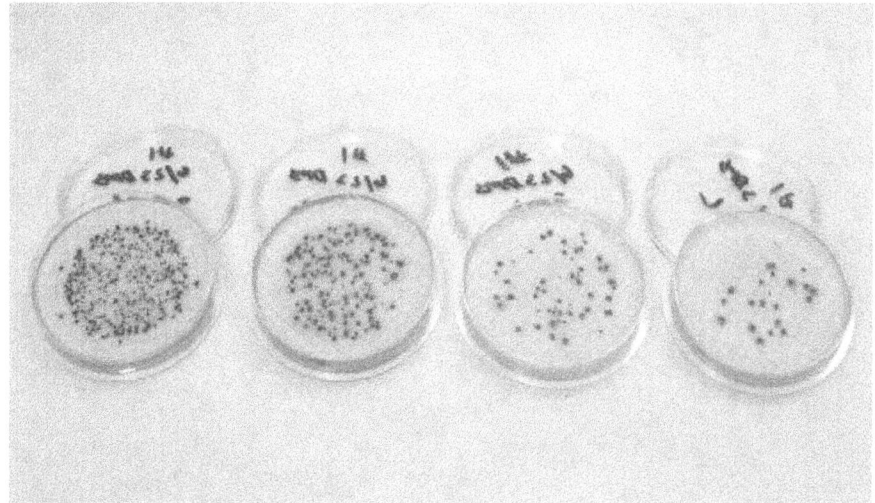

Differing densities of bacteria colonies resulting from various plated sample volumes, modified mTEC method. Magenta colonies are counted as presumptive *E. coli*. (Photo by Donna Francy, U.S. Geological Survey.)

Back to page 3

Example 3

For Huntington, field and laboratory protocols were distributed by the USGS to all personnel. The USGS did several onsite field and laboratory QA/QC checks each year, and corrective actions were taken as needed. Quality-control samples were routinely analyzed during 2004 and 2005, including field blanks and replicate samples for *E. coli*. The laboratory analyzed positive-control reference cultures that were pure cultures of *E. coli* ATCC 10798 (American Type Culture Collection, Rockville, Md.) prepared by the USGS and distributed to field personnel by overnight mail. At the same time, personnel in the USGS laboratory plated the pure culture, and results were compared. For duplicate turbidity measurements, those measurements >10 NTRU that did not agree within 10 percent or <10 NTRU that did not agree within 1 NTRU were repeated. Turbidity reference standards were sent once in 2004 and twice in 2005 to the analyst; corrective actions were taken if results were greater than 25 percent different from standard values.

Turbidimeter and water samples. Turbidities, in nephelometric turbidity ratio units (NTRU), are below the samples. For most samples at Huntington Reservation, turbidity was less than 100 NTRU. (Photos provided by Stephen Lawrence, U.S. Geological Survey.)

Back to page 3

Example 4

Environmental and water-quality data for predictive model development at Huntington were collected by field technicians or compiled from other sources. Field technicians counted the number of birds on the beach upon arrival and estimated wave-height categories at the time of sample collection. Wave heights were placed into four categories based on minimum and maximum heights in each wave train: (1) 0 to 2 ft, (2) 1 to 3 ft, (3) 2 to 4 ft, and (4) > 3 to 5 ft. Wave heights in 2005 were also measured using a second, more accurate method: A survey rod was placed in the water at the sampling location for 1 minute, during which field crews noted the minimum and maximum heights. Lake-level data were obtained from the National Oceanic and Atmospheric Administration (NOAA) station in Cleveland (NOAA ID 9063053) (National Oceanic and Atmospheric Administration, 2005a).

Rainfall and wind-direction data were compiled from the National Weather Service local climatology data station at Hopkins International Airport (National Oceanic and Atmospheric Administration, 2005b). "R_{d-1}" was the amount of rain, in inches, that fell in the 24-hour period (9 a.m. to 9 a.m.) preceding the morning sampling. Similarly, "R_{d-2}" and "R_{d-3}" were amounts of rain that fell in 24-hour periods 2 days and 3 days, respectively, preceding the morning sampling. Weighted rainfall variables were calculated so that the most recent rainfall received the highest weight, as follows:

$$Rw72 = (3 * R_{d-1} + 2 * R_{d-2} + R_{d-3})$$

$$Rw48 = (2 * R_{d-1} + R_{d-2})$$

"Wind direction 24" was calculated by vector addition of hourly wind directions and wind speeds for the 24-hour period preceding sampling (table 1). A vector is a quantity that has both magnitude and direction. A discussion of trigonometric functions and vector math is beyond the scope of this report; a detailed description is provided in Finney and Thomas (1989). Wind directions were then placed into categories by examining patterns in graphs of *E. coli* concentrations and wind direction 24; processes affecting *E. coli* were also considered to ensure the wind direction 24 categories could be explained and seemed reasonable. For example, if one suspected a source of fecal contamination was west of the beach, higher *E. coli* concentrations associated with easterly wind directions would seem reasonable.

(Continued on next page)

Example 4—Continued

Table 1. Example of computing "wind direction 24" by vector addition of hourly wind directions and wind speeds for the 24-hour period preceding sampling for Huntington.

[mi/h, miles per hour; --, not applicable]

| Date and time | Wind direction (degrees) | Quadrant | | | | Wind speed (mi/h) | Cumulative | | Resultant vector magnitude (mi/h) | Resultant vector direction (degrees) |
		I	II	III	IV		X	Y		
05/01/2005 9:53	280	0	0	0	1	15	-14.772	2.605	--	--
05/01/2005 10:53	280	0	0	0	1	15	-29.544	5.209	--	--
05/01/2005 11:53	240	0	0	1	0	11	-39.071	-0.291	--	--
05/01/2005 12:53	280	0	0	0	1	11	-49.903	1.620	--	--
05/01/2005 13:53	250	0	0	1	0	10	-59.300	-1.801	--	--
05/01/2005 14:53	300	0	0	0	1	12	-69.693	4.199	--	--
05/01/2005 15:53	280	0	0	0	1	10	-79.541	5.936	--	--
05/01/2005 16:53	280	0	0	0	1	14	-93.328	8.367	--	--
05/01/2005 17:53	190	0	0	1	0	10	-95.064	-1.481	--	--
05/01/2005 18:53	220	0	0	1	0	14	-104.064	-12.206	--	--
05/01/2005 19:53	200	0	0	1	0	9	-107.142	-20.663	--	--
05/01/2005 20:53	250	0	0	1	0	5	-111.840	-22.373	--	--
05/01/2005 21:53	250	0	0	1	0	3	-114.659	-23.399	--	--
05/01/2005 22:53	310	0	0	0	1	10	-122.320	-16.971	--	--
05/01/2005 23:53	270	0	0	1	0	5	-127.320	-16.971	--	--
05/02/2005 0:53	220	0	0	1	0	6	-131.176	-21.568	--	--
05/02/2005 1:53	230	0	0	1	0	7	-136.539	-26.067	--	--
05/02/2005 2:53	230	0	0	1	0	8	-142.667	-31.209	--	--
05/02/2005 3:53	220	0	0	1	0	6	-146.524	-35.806	--	--
05/02/2005 4:53	220	0	0	1	0	8	-151.666	-41.934	--	--
05/02/2005 5:53	230	0	0	1	0	6	-156.262	-45.791	--	--
05/02/2005 6:53	240	0	0	1	0	8	-163.191	-49.791	--	--
05/02/2005 7:53	250	0	0	1	0	13	-175.407	-54.237	--	--
05/02/2005 8:53	270	0	0	1	0	13	-188.407	-54.237	196	254
							0	0		

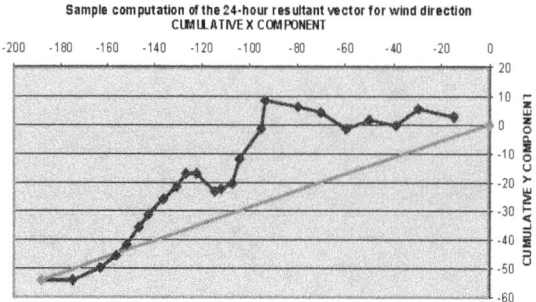

Sample computation of the 24-hour resultant vector for wind direction
CUMULATIVE X COMPONENT

Back to page 4

Example 5

In Ohio, 235 colony forming units per 100 milliliter (CFU/100 mL) is used as a single-sample maximum bathing-water standard (Ohio Environmental Protection Agency, 2003) for beach notification and closure decisions, effective December 2005 (U.S. Environmental Protection Agency, 2004). Although the geometric mean of 126 CFU/100 mL was used in Ohio before December 2005, the single-sample maximum was used as a benchmark to evaluate water quality and model performance at Huntington in this report. Median annual concentrations of *E. coli* at Huntington ranged from 34 to 110 CFU/100 mL for the 6 years of this study (table 2). The percentage of days that the single-sample maximum bathing-water standard was exceeded ranged from 11.1 percent in 2003 to 23.5 percent in 2000.

Table 2. Summary statistics of *Escherichia coli* (*E. coli*) concentrations at Huntington, 2000–2005.

[CFU/100 mL is colony-forming units per 100 milliliters]

Year	Number of samples	Daily *E. coli* concentration[a] (CFU/100 mL)			Number (percent) of days bathing-water standard was exceeded[b]
		Median	Minimum	Maximum	
2000	51	110	8	6,600	12 (23.5)
2001	50	44	3	1,200	10 (20.0)
2002	52	43	4	1,800	11 (21.2)
2003	54	58	2	730	6 (11.1)
2004	54	31	3	1,500	7 (13.0)
2005	58	34	1	2,400	8 (13.8)

[a] The daily concentrations of *E. coli* were determined by calculating the average of two or three point samples.

[b] Number of days the concentration of *E. coli* in water exceeded the single-sample maximum bathing-water standard of 235 CFU/100 mL.

Back to page 4

Example 6

Two example scatterplots of the Huntington 2000–2004 data are shown in figure 2. Turbidity shows a positive linear relation to *E. coli* (fig. 2, left-hand graph). No extreme values were shown to appreciably influence the relation between *E. coli* and turbidity. Examining this relation by year in scatterplots confirmed a consistent relation from year to year (data not shown). In contrast, the relations between *E. coli* and day of the year differed from year to year (fig. 2, right-hand graph): (a) triangles for 2000 show a negative relation, (b) circles for 2001 and 2002 show a positive relation, and (c) squares for 2003 and 2004 show a nearly horizontal line indicating no relation. The differences between each year decreased the usefulness of this variable for predictive purposes. In addition, the relation for 2000 was strongly influenced by three extreme values that were greater than 2,000 CFU/100 mL and were associated with high amounts of rainfall that spring.

Categorical data at Huntington for 2000–2004 also were graphically examined. *E. coli* concentrations increased with increasing wave height (fig. 3, left-hand graph). The median *E. coli* concentration for wave-height category 3–6 ft exceeded the single-sample maximum bathing-water standard of 235 CFU/100 mL and was just below the standard for wave height category 2–4 ft. In contrast, differences in *E. coli* concentrations among 24-hour wind-direction categories were not evident (fig. 3, right-hand graph).

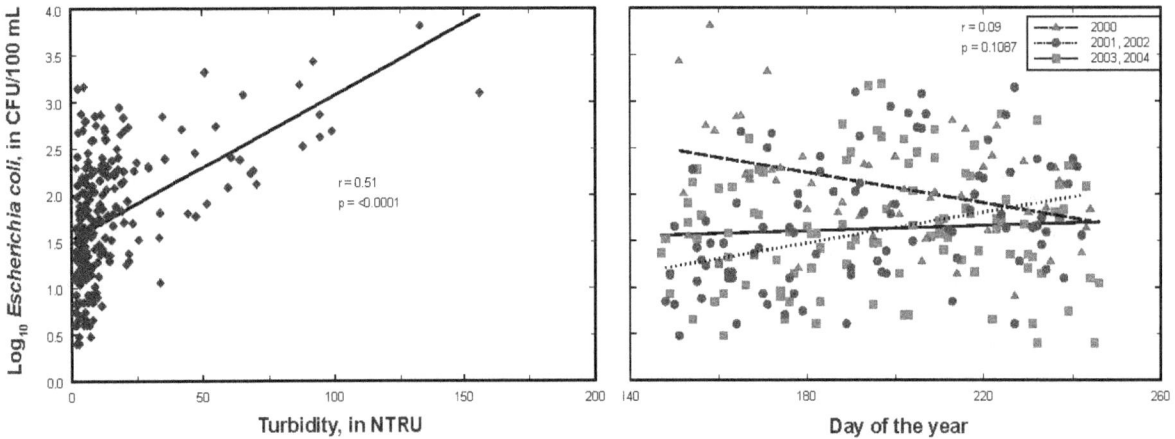

Figure 2.   Huntington, 2000–2004, relations between *Escherichia coli* concentrations and turbidity and day of the year. (r is the correlation coefficient, and p is the significance of the correlation for all years combined; least-square regression lines for each year are included. CFU/100 mL is colony-forming units per 100 milliliters; NTRU is nephelometric turbidity ratio units.)

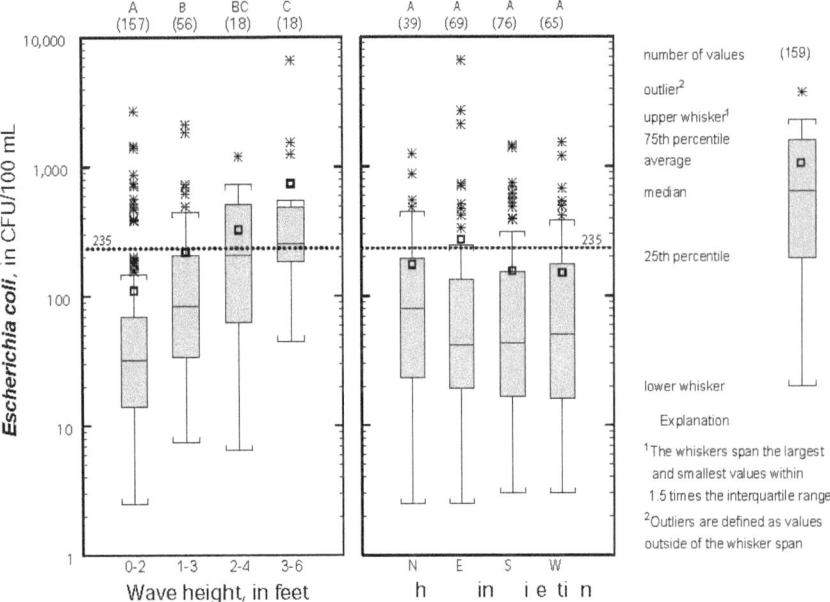

Figure 3.   Huntington, 2000–2004, *Escherichia coli* concentrations in water, by wave height and 24-hour wind direction. (Results of Tukey's test are presented as letters; concentrations with at least one letter in common do not differ significantly. The Ohio single-sample maximum bathing-water standard of 235 CFU/100 mL is indicated by dotted lines and used as a benchmark. CFU/100 mL is colony-forming units per 100 milliliters.)

Back to page 4

Example 7

Correlations between *E. coli* concentrations and potential explanatory variables for data collected during 2000–2004 are shown in aggregate and by year for Huntington (table 3, left side of the solid line). R_{d-1}, turbidity, and $\log_{10}$ turbidity were positively and significantly correlated to *E. coli* for all 5 years. The variable with the strongest correlation to *E. coli* for all years combined was log turbidity (r = 0.54). Day of the year, R_{d-2}, water temperature, and lake level were significantly correlated to *E. coli* during some years, but not all. Day of the year and water temperature showed significant correlations that were positive during 1 or 2 years and negative during 1 year. Because both R_{d-1} and R_{d-2} were significantly correlated to *E. coli* for all years combined, Rw48 was included as an additional variable. Rw48 improved the correlations to *E. coli* over single-day rainfall variables for all single years except for 2001 and improved the correlations for 2000–2004 combined. R_{d-3} and number of birds were not positive and significantly correlated to *E. coli* during any of the years tested. Although number of birds was significantly correlated to *E. coli* during 2000, the correlation was negative; this does not seem reasonable, because one would expect higher *E. coli* concentrations with higher bird numbers.

Table 3. Pearson's r correlations between $\log_{10}$ *Escherichia coli* concentrations and explanatory variables for Huntington, 2000–2005.

[Relations that were significant at p< 0.05 are in bold and italic]

Variables	2000	2001	2002	2003	2004	2000-2004	2005	2000-2005
Birds, number at time of sampling	*-0.35*	0.12	-0.17	-0.16	-0.09	-0.10	0.20	0.03
Day of the year	*-0.38*	*0.35*	*0.35*	0.02	0.14	0.09	*0.36*	*0.15*
R_{d-1} [a]	*0.47*	*0.27*	*0.24*	*0.36*	*0.50*	*0.34*	*0.44*	*0.36*
R_{d-2} [a]	*0.28*	-0.06	-0.02	0.20	*0.27*	*0.20*	*0.32*	*0.22*
R_{d-3} [a]	0.03	-0.04	-0.12	0.10	-0.09	0.08	0.13	0.08
Rw48 [b]	*0.50*	0.23	*0.28*	*0.38*	*0.55*	*0.37*	*0.53*	*0.40*
Turbidity	*0.60*	*0.50*	*0.49*	*0.35*	*0.51*	*0.51*	*0.38*	*0.48*
$\log_{10}$ turbidity	*0.63*	*0.58*	*0.38*	*0.49*	*0.54*	*0.54*	*0.40*	*0.51*
Water temperature	*-0.36*	*0.51*	0.23	-0.06	0.26	*0.13*	-0.23	0.002
Lake level	-0.12	-0.15	*-0.30*	0.05	0.09	-0.11	*-0.34*	*-0.16*
Wave-height rod	--	--	--	--	--	--	*0.48*	--

[a] R_{d-1} was the rainfall amount, in inches, at Hopkins International Airport, Cleveland, Ohio, in the 24-hour period preceding sampling; R_{d-2} and R_{d-3} were the rainfall amounts 2 and 3 days, respectively, before sampling.

[b] Rw48 was the rainfall amount, in inches, at Hopkins International Airport, Cleveland, Ohio, in the 48-hour period before sampling, with the most recent rainfall receiving the most weight.

(Continued on next page)

Example 7—Continued

The relations among explanatory variables at Huntington for 2000–2004 data were also examined (table 4). Strong significant correlations were found for Rw48 and the two single-day rainfall variables; this is not surprising because R_{d-1} and R_{d-2} are components of Rw48. Day of the year and water temperature also were strongly correlated. Combining these variables into a model may cause problems with collinearity. Weaker, but significant correlations were those between turbidity and rainfall variables or date. The correlation between day of the year and lake level was negative and significant, indicating that, as the summer progressed, lake levels decreased. The variables that were weakly correlated will probably not cause problems with collinearity in the model.

Results of ANOVA and Tukey's test for data collected 2000–2004 on two categorical variables—wave height and wind direction—are shown in figure 3. Statistically significant differences in *E. coli* concentrations were found between wave-height category 0 to 2 ft and all other categories; wave-height category 3 to 6 ft differed significantly from the two lowest categories, but not from the 2 to 4 ft category (fig. 3, example 6). No statistically significant differences were found in *E. coli* concentrations among 24-hour wind direction categories (fig. 3, example 6).

Table 4.   Pearson's r correlations among explanatory variables for Huntington, 2000–2004.

[Relations that were significant at $p < 0.05$ are in bold and italic]

	R_{d-1}[a]	R_{d-2}[a]	Rw48[b]	Turbidity	Log turbidity	Water temperature	Day of the year
R_{d-2}	*0.18*	--	--	--	--	--	--
Rw48	*0.89*	*0.61*	--	--	--	--	--
Turbidity	*0.18*	*0.18*	*0.23*	--	--	--	--
Log_{10} turbidity	*0.14*	*0.15*	*0.18*	*0.83*	--	--	--
Water temperature	-0.06	*-0.13*	-0.11	*-0.16*	*-0.28*	--	--
Day of the year	-0.07	-0.10	-0.10	*-0.14*	*0.23*	*0.70*	--
Lake level	0.07	0.04	0.08	0.01	0.08	*-0.13*	*-0.20*

[a] R_{d-1} was the rainfall amount, in inches, at Hopkins International Airport, Cleveland, Ohio, in the 24-hour period preceding sampling; R_{d-2} was the rainfall amount 2 days before sampling.

[b] Rw48 was the rainfall amount, in inches, at Hopkins International Airport, Cleveland, Ohio, in the 48-hour period before sampling, with the most recent rainfall receiving the most weight.

Back to page 4

Example 8

The following variables showed consistent significant relations with *E. coli* for 2000–2004 and were used to develop a list of models for Huntington: wave height, R_{d-1}, Rw48, turbidity, and log turbidity. The resultant possible models, along with their Mallow's Cp statistic and R^2 values, are listed in table 5. For Huntington 2000–2004, models that included turbidity and log turbidity together, or Rw48 and R_{d-1} together, were not considered because of potential problems with collinearity. Use of single variables produced models with low R^2 values ranging from 0.11 for R_{d-1} to 0.29 for $\log_{10}$ turbidity and high Mallows' Cp statistics. Model 1—wave height, Rw48, and $\log_{10}$ turbidity—had the lowest Mallows' Cp statistic (3.240) and the highest R^2 (0.38) among all the models. The next model on the list without related variables, model 4, had a slightly higher Mallows' Cp statistic than model 1. Model 1, however, was chosen for further testing because it included two days of rainfall instead of one; from the exploratory data analysis, two days of weighted rainfall improved the relation to *E. coli* over one day of rainfall. The equation for model 1 (Huntington 2000–2004 model) is as follows:

$$\text{Log}_{10}\,(E.\,coli) = 0.144*\text{wave height} + 0.301*\text{Rw48} + 0.563*\log_{10}\text{turbidity} + 0.914$$

Table 5.   List of possible models and the Mallows' Cp test for Huntington, 2000–2004.

[R^2 is the coefficient of determination. The Cp statistic (Mallows, 1973) is a measure of the error in a model with a subset of explanatory variables, relative to the error in a model that incorporates all potential explanatory variables. Log turbidity is $\log_{10}$ turbidity. R_{d-1} was the rainfall amount, in inches, at Hopkins International Airport, Cleveland, Ohio, in the 24-hour period preceding sampling. Rw48 was the rainfall amount, in inches, at Hopkins International Airport, Cleveland, Ohio, in the 48-hour period before sampling, with the most recent rainfall receiving the most weight.]

Model	Number of variables	Cp	Adjusted R^2	Variables in model
1	3	3.240	0.38	Wave height, Rw48, log turbidity
2	4	4.470	0.38	Wave height, Rw48, log turbidity, turbidity
3	4	4.802	0.38	Wave height, R_{d-1}, Rw48, log turbidity
4	3	5.999	0.38	Wave height, R_{d-1}, log turbidity
5	2	11.200	0.36	Rw48, log turbidity
6	3	12.033	0.36	R_{d-1}, Rw48, log turbidity
7	2	12.290	0.36	R_{d-1}, log turbidity
8	3	12.734	0.36	Wave height, Rw48, turbidity
9	3	16.006	0.35	Wave height, R_{d-1}, turbidity
10	2	28.219	0.32	Rw48 , turbidity
11	2	29.080	0.32	Wave height, log turbidity
12	2	29.440	0.32	R_{d-1}, turbidity
13	2	34.349	0.30	Wave height, Rw48
14	2	36.179	0.30	Wave height, turbidity
15	1	39.136	0.29	Log turbidity
16	2	41.965	0.28	Wave height, R_{d-1}
17	1	53.177	0.25	Turbidity
18	1	67.874	0.22	Wave height
19	1	101.64	0.13	Rw48
20	1	109.06	0.11	R_{d-1}

Back to page 5

Example 9

Regression diagnostics were done on the Huntington 2000–2004 model. The parameter estimates were reasonable in value and significant (table 6), and no observation was found to have a Cook's D above the critical value (data not shown). The partial residual plots showed patterns of a general increase in each of the explanatory variables with increases in *E. coli* (fig. 4). Plots of the residuals versus predicted values showed that there were generally constant variances throughout the data sets; although residuals were smaller for higher predictive values, this was not a concern because there were fewer observations at the high end than at the low end (fig. 5). A plot of measured and predicted *E. coli* concentrations (fig. 6) showed that the relation was linear, although there was considerable error in the predicted values.

Table 6. Huntington 2000–2004 model, statistics and parameter estimates.

[Log turbidity is $\log_{10}$ turbidity. Rw48 was the rainfall amount, in inches, at Hopkins International Airport, Cleveland, Ohio, in the 48-hour period before sampling, with the most recent rainfall receiving the most weight]

Model 1 regression Dependent variable: Log *Escherichia coli*					
Source	**Degrees of freedom**	**Sum of squares**	**Mean square**	**F value**	**P value**
Model	3	42.48	14.16	52.45[a]	<0.0001
Error	244	65.87	0.27		
Corrected total	247	108.35			
	Root mean square error	0.52	R-squared	0.39	
	Dependent	1.74	Adjusted R-squared	0.38[b]	
	Coefficient variance	29.77			

Parameter estimates					
Variable	**Degrees of freedom**	**Parameter estimate**	**Standard error**	**t value**	**P value**
Intercept	1	0.914	0.079	11.560	<0.0001[c]
Wave height	1	0.144	0.046	3.160	0.0018[c]
Rw48	1	0.301	0.057	5.280	<0.0001[c]
Log turbidity	1	0.563	0	5.760	<0.0001[c]

[a] The F-test indicates that the regression model explains a significant proportion of the variance in the dependant variable.

[b] The adjusted R-squared indicates the fraction of the variation in *E. coli* concentration explained by the model, adjusted for number of variables in the model.

[c] P values of t-tests indicate that each of the parameter estimates are statistically different from zero.

(Continued on next page)

Example 9—Continued

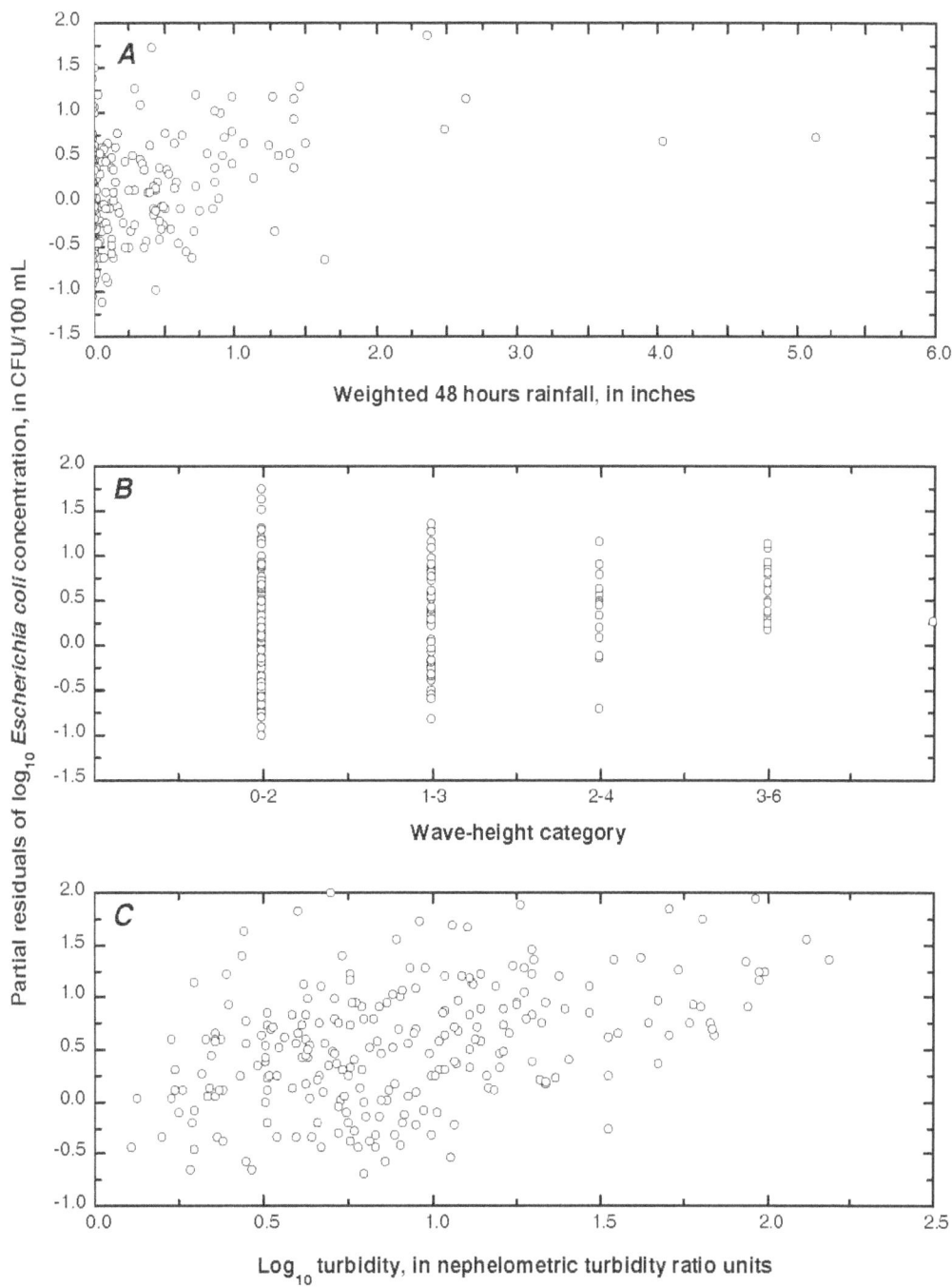

Figure 4.   Partial residual plots of explanatory variables for the Huntington 2000–2004 model. *A*, Weighted 48-hour rainfall. *B*, Wave height. *C*, Log$_{10}$ turbidity. (CFU/100 mL is colony-forming units per 100 milliliters.)

(Continued on next page)

Example 9—Continued

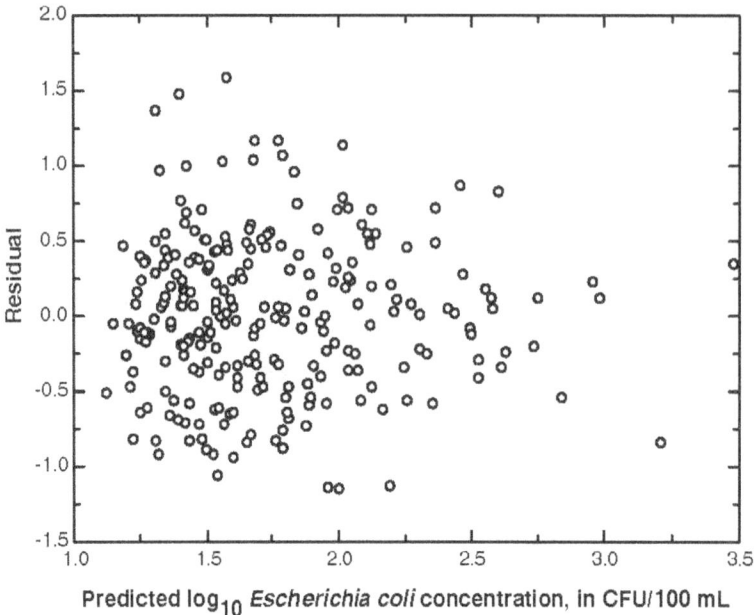

Figure 5. Predicted *Escherichia coli* concentrations and residuals for the Huntington 2000–2004 model. (CFU/100 mL is colony-forming units per 100 milliliters.)

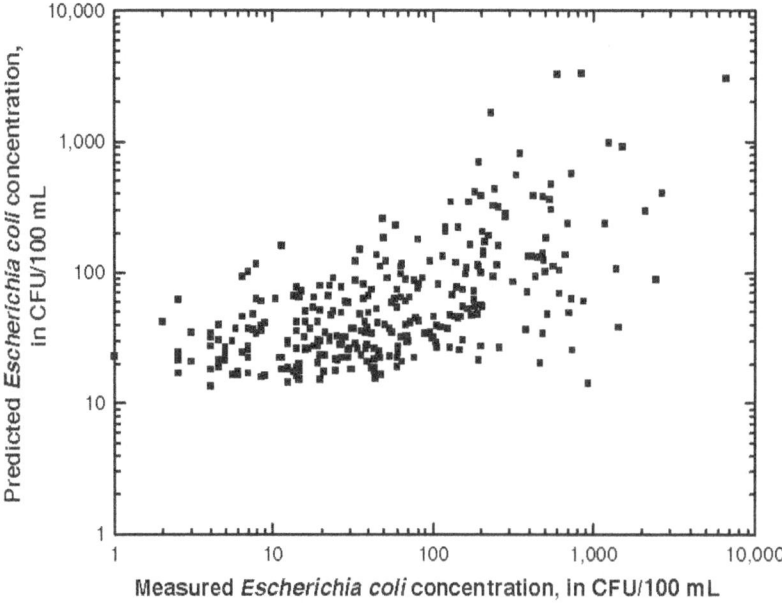

Figure 6. Measured and predicted *Escherichia coli* concentrations for the Huntington 2000–2004 model. (CFU/100 mL is colony-forming units per 100 milliliters.)

Back to page 6

Example 10

The Huntington 2000–2004 model was used to predict output values for the data used to develop the model. Predicted *E. coli* concentrations were output by the statistical software. Determining exceedance probabilities require further calculations; the Fortran program to make these calculations is included in Appendix 1, Example 1.4.

When analyzing predicted *E. coli* concentrations as output values, calculation of a target value is not needed because the target is, by default, the bathing-water standard. When analyzing exceedance as a probability, a threshold probability must be determined—the lowest (most conservative) probability that produces the most correct responses and (or) fewest false negative responses (Francy and others, 2003). This concept can be best explained by examining the plot for the Huntington 2000–2004 best model with a 29-percent threshold (fig. 7) and then explaining the process used to determine the 29-percent threshold. The plot is divided into four quadrants by a vertical line through 235 CFU/100 mL on the x-axis and a horizontal line through threshold probability of 29. The four quadrants are

- **Correct nonexceedance.** *E. coli* concentration met the standard (was less than 235 CFU/100 mL), and the predicted probability of exceedance was below the threshold.

- **False positive.** *E. coli* concentration met the standard, but the predicted probability of exceedance was above the threshold.

- **Correct exceedance.** *E. coli* concentration exceeded the standard, and the predicted probability of exceedance was above the threshold.

- **False negative.** *E. coli* concentration exceeded the standard, but the predicted probability of exceedance was below the threshold.

By raising or lowering the horizontal line, one can determine the best threshold probability. This determination is somewhat subjective. For example, a threshold of 50 would have produced the highest number of correct responses (215) but would also have produced a high number of false negatives (28). False negative responses are especially troubling because the recreational water quality is determined to be acceptable when in fact the standard was exceeded. Thresholds between 35 and 45 do little to reduce the number of false negatives. Selecting a threshold of 29, however, still maintains a high number of correct responses (210) but yet reduces the false negatives to a more acceptable level (18) and represents a compromise between false negative and false positive responses. In addition, setting the threshold to a lower value such as 29 enables the beach manager to err on the safe side.

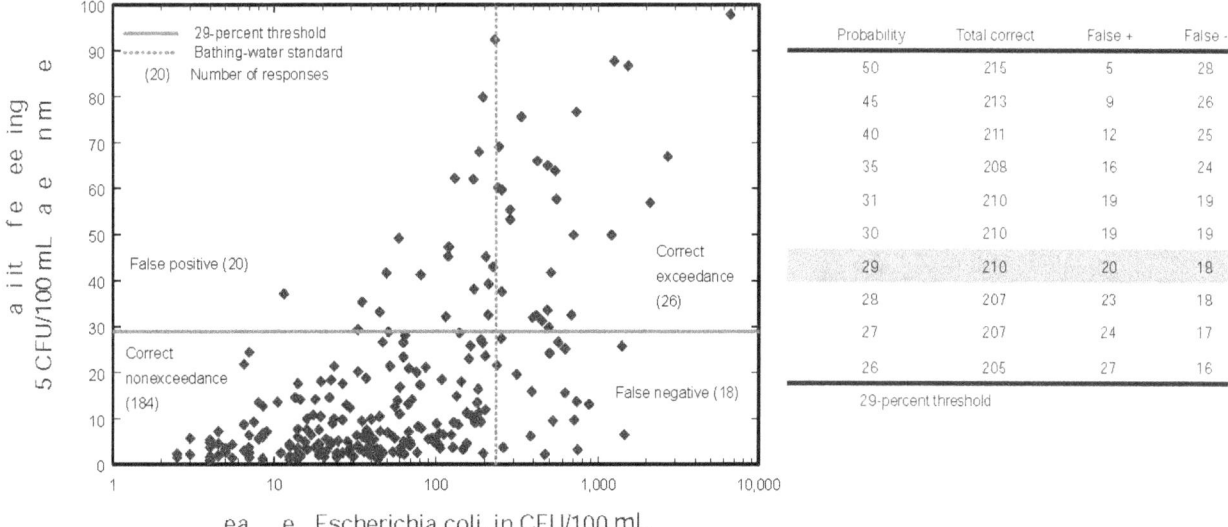

Figure 7. Establishment of the threshold probability for the Huntington 2000–2004 model. (CFU/100 mL is colony-forming units per 100 milliliters.)

(Continued on next page)

Example 10—Continued

Model responses were compared to use of the previous day's *E. coli* concentration (table 7). For the Huntington 2000–2004 model, the percentages of correct predictions, specificities, and sensitivities were higher using the model than using the previous day's *E. coli* concentration. Model specificities were relatively high for probability and predicted *E. coli* outputs (90.2 and 97.5 percent, respectively), but model sensitivities were lower (59.1 and 36.4 percent, respectively). When the standard was exceeded (sensitivity), use of threshold probabilities resulted in a better response than use of the predicted *E. coli* or previous day's *E. coli*.

Table 7. Huntington, numbers of correct responses and the sensitivities and specificities of model responses with indicated thresholds and predicted *Escherichia coli* (*E. coli*) concentrations compared to previous day's *E. coli* concentrations (current method for assessing recreational water quality).

Beach model years	Threshold probability	Number of samples	Response (percent)		
			Correct predictions	Specificity[a]	Sensitivity[b]
2000–2004	29	248	84.7	90.2	59.1
	Predicted *E. coli*	248	86.7	97.5	36.4
	Previous day's *E. coli*	171	76.6	86.5	30.0
2000–2005	27	306	85.9	90.9	61.5
	Predicted *E. coli*	306	85.6	96.4	32.7
	Previous day's *E. coli*	213	76.5	87.0	25.0

[a] Specificity was the proportion of nonexceedance responses that were correctly predicted as safe for swimming.

[b] Sensitivity was the proportion of exceedance responses that were correctly predicted as unsafe for swimming.

Back to page 6

Example 11

The Huntington 2000–2004 model with the 29-percent threshold was validated in 2005, and model responses were compared to using the previous day's *E. coli* concentrations (fig. 8). The percentage of correct predictions using the previous day's *E. coli* concentrations (75.6 percent) was lower than using either model output (82.0 and 88.0 percent). The specificities found using the model outputs and the previous day's *E. coli* concentration were in the same range (88.1 to 95.2 percent); however, using the previous day's *E. coli* concentration resulted in fewer total predictions (41) than the model (50) because samples were not collected on Sundays. The specificity was slightly higher using the predicted *E. coli* (95.2 percent) than the probability (88.1 percent) as the model output variable. The difference between the model responses and the current method response is most pronounced for sensitivities. Using either output values from the model, 4 out of 8 exceedances during 2005 were correctly predicted (50 percent sensitivity). Using the previous day's *E. coli*, none of the exceedances were predicted, resulting in a sensitivity of zero.

The data collected at Huntington during 2005 were added to the 2000–2004 dataset, and the model-development process was followed with the additional year of data. Correlation coefficients that describe the relations between explanatory variables and *E. coli* for 2005 and for 2000–2005 combined are listed in table 3 (right side of the solid line) for a comparison to earlier years. As in 2000–2004, the relations between *E. coli* and R_{d-1}, R_{d-2}, Rw48, turbidity, and $\log_{10}$ turbidity were significant for the 2000–2005 dataset. With the additional year, day of the year and lake level were significantly related to *E. coli* for 2000–2005 and were, therefore, added as possible explanatory variables during the 2000–2005 model-development process. Wave height measured with a survey rod was significantly related to *E. coli* during 2005 but was not used in the model because only 1 year of wave-height data were collected in this manner.

A list of possible models was developed for Huntington based on 2000–2005 data along with the Mallows' Cp statistic. The best model contained the variables wave height, Rw48, $\log_{10}$ turbidity, and day of the year with an adjusted R^2 of 0.42, an improvement over the R^2 for the Huntington 2000–2004 model. The equation for the Huntington 2000–2005 model is as follows:

$$\text{Log}_{10}\,(E.\,coli) = 0.134 * \text{wave height} + 0.293 * \text{Rw48} + 0.592 * \text{log turbidity} + 0.006 * \text{day of the year} - 0.219$$

The new model passed regression diagnostics and hypothesis tests. The established threshold probability of 27 percent for the Huntington 2000–2005 model yielded similar responses as the 29 percent threshold for the Huntington 2000-2004 model (table 7). The sensitivity for the Huntington 2000–2005 model (61.5 percent) using the threshold probability was considerably higher than the sensitivity achieved using the previous day's *E. coli* (25.0 percent) or the predicted *E. coli* concentration (32.7 percent). The 2000–2005 Huntington model will be validated in 2006 and used as a predictive tool by beach managers.

Continued on next page

Example 11—Continued

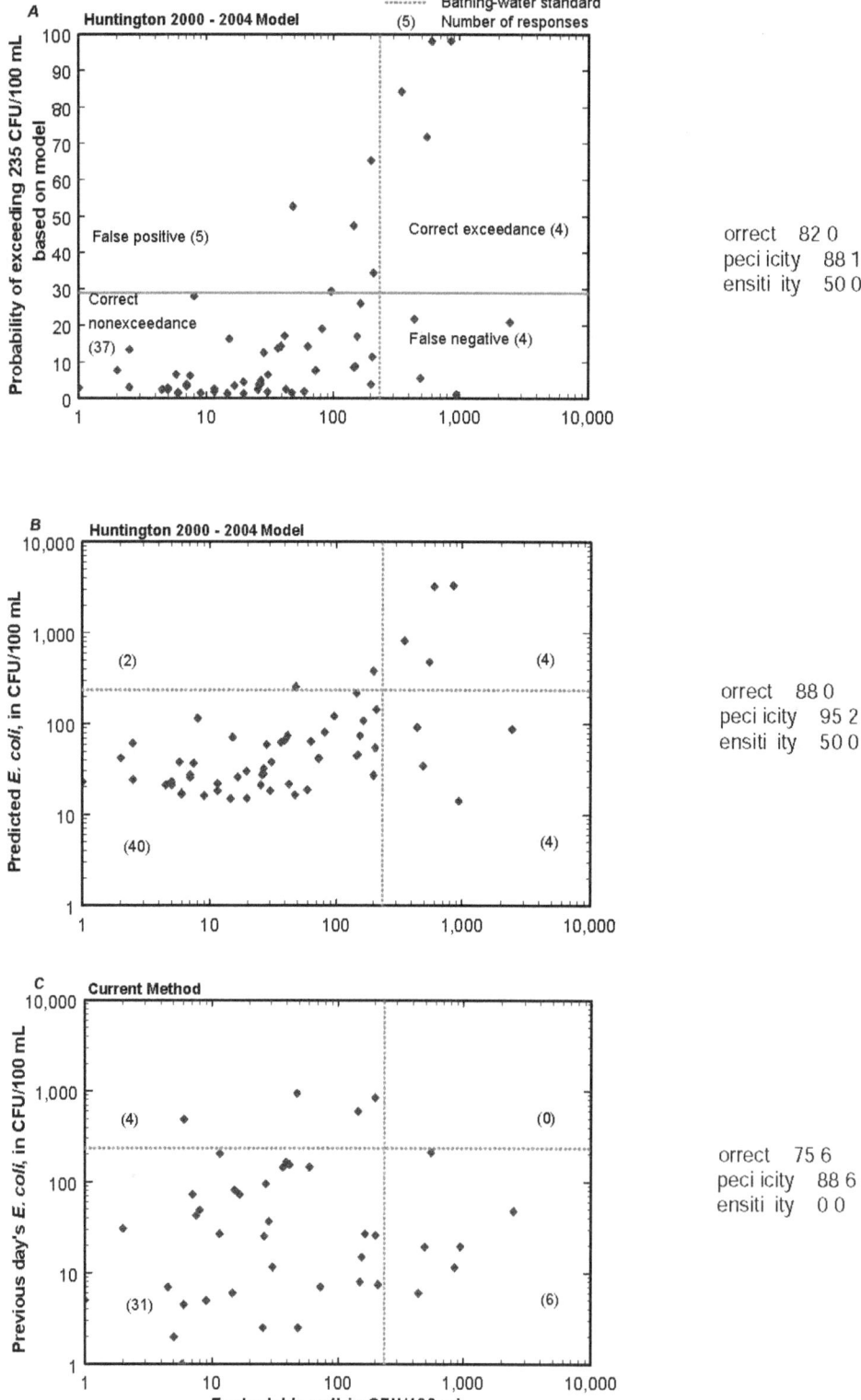

Figure 8. Huntington 2005, performance in assessing recreational water-quality of the Huntington 2000–2004 model. *A*, Probability output, and *B*, predicted *Escherichia coli* (*E. coli*) output compared to *C*, current method. (CFU/100 mL is colony-forming units per 100 milliliters.)

Back to page 6

Example 12

An Internet-based "nowcasting" system was developed for Lake Erie beaches and is being used for Huntington at the date of writing (August 2006) (see *http://www.ohionowcast.info*); the system became operational on May 30, 2006. Recreational water-quality conditions are estimated by means of the Huntington 2000–2005 model and transmitted through the nowcasting system 7 days per week. Advisories are issued if the probability of exceeding the single-sample maximum bathing-water standard exceeds 27 percent.

Future steps to improve the predictive models at Huntington and other Lake Erie beaches include the incorporation of more accurately measured wave heights, continuous turbidity measurements, locally installed rain gages, rapid analytical methods for *E. coli*, hydrodynamic modeling, and weekend sampling.

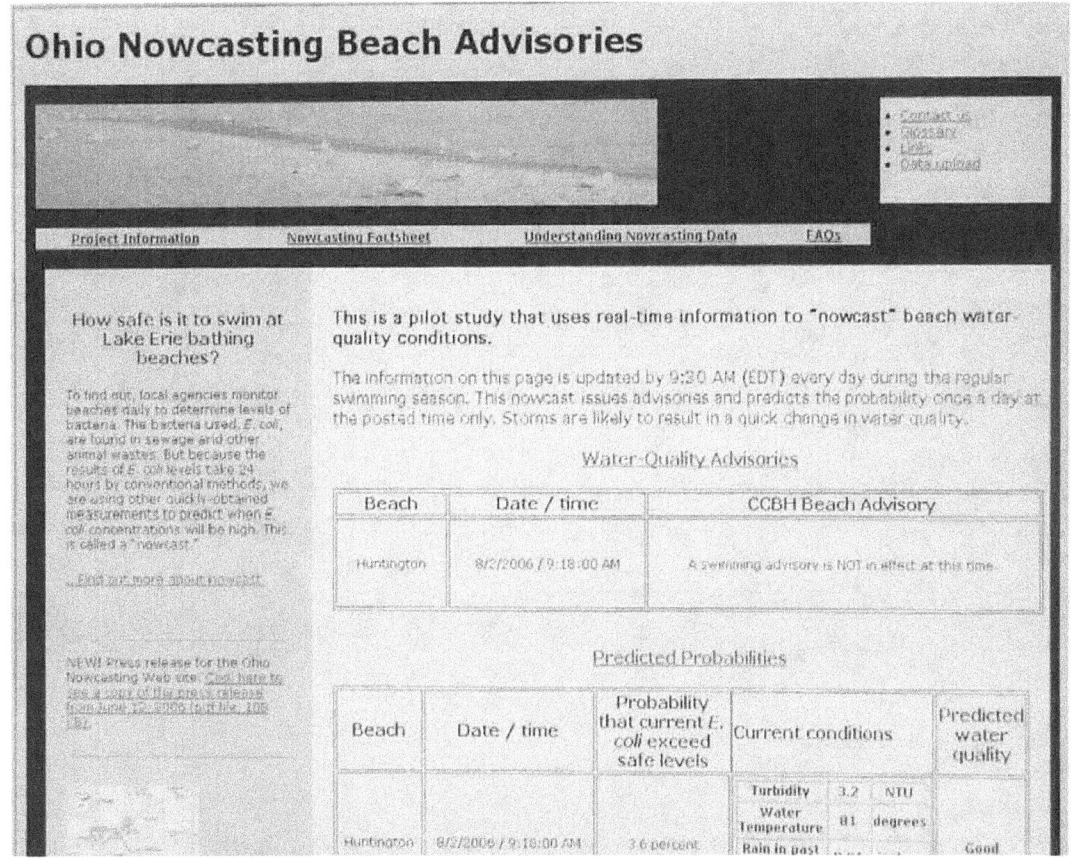

Screen shot of the Ohio Nowcasting Beach Advisories Web page for a day in summer 2006.

Back to page 7

Appendix 1

SAS commands to determine the best 50 models and to obtain individual model parameters and Fortran program to determine the probability of exceeding the single-sample maximum bathing-water standard (235 colony-forming units per 100 milliliters).

The Fortran code was originally developed by Gary Tasker (U.S. Geological Survey, Reston, Virginia, retired) for a report by Francy and Darner (1998). The program below was modified for the Huntington 2000–2004 model.

For the following commands, the SAS data set called "two" contains all data corresponding to the beach and time period referenced in the comment lines. (SAS Institute, Inc., 1990).

Example 1.1. SAS commands to determine the best 50 models based on R^2 and Mallow's Cp. The output from these commands will be similar to that shown in table 5.

```
proc printto file='c:\best50_2000_2004.txt';
options linesize=70 pagesize=52 pageno=1;
title 'Huntington Best Models - 2000-2004';
proc reg data=two;
model logecoli = Turbidity WaveHt precip Step w48 logturb
  / selection = cp best = 50 adjrsq;
output out=resplot p=pred r=resid;
run;
proc printto;run;
title '';run;
```

Example 1.2. SAS commands to obtain results for individual model parameters for model diagnosis and selection and for use in the Fortran program to compute probabilities (example 1-4):

```
proc printto file='c:\mlr.txt';
options linesize=98 pagesize=52 pageno=1;
title 'Huntington MLR - 2000-2004';
proc reg data=two;
model logecoli = waveht w48 logturb
  / r partial covb;
plot ( logecoli)*( waveht w48 logturbl)
 residual.*predicted.
 residual.*( waveht w48 logturb);
output out=resplot p=pred r=resid;
run;
proc printto;run;
title '';run;
```

Example 1-3. Output from SAS for individual model parameters (mlr.txt).

```
                    The REG Procedure
                   Model: MODEL1
                Dependent Variable: logecoli

                  Analysis of Variance

                        Sum of      Mean
      Source        DF  Squares     Square  F Value  Pr > F

      Model          3  42.48015  14.16005   52.45   <.0001
      Error        244  65.87395   0.26998
      Corrected Total 247 108.35410


         Root MSE                  R-Square   0.3920
         Dependent Mean   1.74555  Adj R-Sq   0.3846
         Coeff Var       29.76664


                  Parameter Estimates

                    Parameter    Standard
        Variable  DF  Estimate     Error   t Value  Pr > |t|

        Intercept  1   0.91432    0.07906   11.56    <.0001
        WaveHt     1   0.14416    0.04556    3.16    0.0018
        w48        1   0.30089    0.05694    5.28    <.0001
        logturb    1   0.56274    0.09765    5.76    <.0001


                  Covariance of Estimates

      Variable    Intercept       WaveHt        w48         logturb

      Intercept  0.0062507263  -0.000810778  -0.000171345   -0.00420649
      WaveHt     -0.000810778   0.002076132  -0.000151048   -0.002687444
      w48        -0.000171345  -0.000151048   0.0032421405  -0.000624654
      logturb    -0.00420649   -0.002687444  -0.000624654    0.0095355972
```

Example 1-4. Fortran program to determine the probability of exceeding the single-sample maximum bathing-water standard (235 colony-forming units per 100 milliliters) The output from SAS in example 1.3 has been color coded to match parameters in the Fortran code.

```
c ===============================================================
c program to compute e.coli concentrations at Huntington Beach
c using data from recreational seasons 2000 through 2004
c Rainfall at Hopkins international Airport,
c Wave heights at the beach, turbidity at the beach

c np= number of parameters, including the intercept)
c t90 = t for number of degrees of freedom in the model and alpha = 0.05
c ndf = number of degrees of freedom in regression
c xtx = covariance matrix
c b  = vector of regression coefficients
c v  = vector of explanatory variable
c vt = transpose of b

   parameter (np=4)
   character*16 name
   real xtx,v,vt,b,temp,var,turb,w48,waveht,rain1,rain2,lturb
   dimension xtx(np,np),v(1,np), vt(np,1), b(np),
   & temp(1,np),var(1,1)
   data name /'Huntington  '/
   data ndf /247/
   data rmse /0.51959/, t90 /1.645/
   data b/.91432,.14416,.30089,.56274/
c
    data xtx /0.0062507263,-0.000810778,-0.000171345,-0.00420649,
   1 -0.000810778,0.002076132,-0.000151048,-0.002687444,
   1 -0.000171345,-0.000151048,0.0032421405,-0.000624654,
   1 -0.00420649,-0.002687444,-0.000624654,0.0095355972/


c
c Initialize variables
c
   print *, ' Program computes E. coli concentrations at'
   print *, ' Huntington Beach            '
   print *, ' Estimates are based on multiple linear regressions'
   print *, ' of 2001-2004 data'
   print *, ' Last compiled on March 3, 2005'
   print *, ' ****************************************************'
c
c Enter wave height data
c
c
   print *, ' ENTER wave height index'
   print *, ' Enter               FT  '
   print *, '  1   if wave height is between   0 and 2'
   print *, '  2   if wave height is between   1 and 3'
   print *, '  3   if wave height is between   2 and 4'
   print *, '  4   if wave height is between   3 and 6'
   write (*,*)
   write (*,*)
   read (*,*) waveht


c
c Enter turbidity data
c
   print *, ' ENTER Turbidity at Huntington'
   print *, ' Turbidity in NTUs= '
   read (*,*) turb
   lturb = log10(turb)
```

```fortran
c
c Enter rainfall data
c
   print *, ' ENTER rainfall at Hopkins, in inches for the last'
   print *, ' two days. '
   print *, ' Rain two days ago ='
   read (*,*) rain2
   print *, ' Rain in past 24 hours ='
   read (*,*) rain1
   w48 = (2.0*rain1)+rain2

c
c Set vector elements
c
   v(1,1)=1.0
   v(1,2)=waveht
   v(1,3)=w48
   v(1,4)=lturb
c***
   write (*,222)waveht,w48,lturb
 222 format(' waveht,w48,lturb ',4f10.3)
c***
c
c
   write(*,100)
c
c Transpose vector v
c
    do 2 i=1,np
  2  vt(i,1)=v(1,i)
c
c Compute log of bacteria from linear regression model
c
   y= b(1)
   do 3 ip=2,np
    y=y+v(1,ip)*b(ip)
  3 continue
   ecol=10.0**y
c
c Compute standard error of a new prediction
c
   call mltply(temp,v,xtx,1,np,np,1,1,np)
   call mltply(var,temp,vt,1,np,1,1,1,np)
   vp=rmse**2+var(1,1)
   sep=sqrt(vp)
c
c Compute probability of exceeding 235
c
   t=(alog10(235.0)-y)/sep
   probt=1.0-stutp(t,ndf)
c
c Compute 90 percent prediction interval
c
   cl90=y-t90*sep
   cu90=y+t90*sep
   cl90=10.0**cl90
   cu90=10.0**cu90
c
c Write results
c
   write (*,200) ecol, cl90, cu90, probt
  1 continue
 200 format(14x,3f10.1,5x,f10.3)
 100 format(t2,' Huntington','   E. coli ',' | 90% pred. int.|',
   & '|   Prob>235  |',/,t27,' | lower  upper |',
   & '|             |',/)
```

```
C
   stop
   end
C
   SUBROUTINE MLTPLY(PROD,X,Y,K1,K2,K3,N1,N2,N3)
C    IMPLICIT REAL*8 (A-H,O-Z)
   INTEGER i, j, k, K1, K2, K3, N1, N2, N3
   REAL PROD(N1,K3), X(N2,K2), Y(N3,K3), sum
C ------------------------------------------------------------
C X IS K1*K2 MATRIX
C Y IS K2*K3 MATRIX
C PROD = X*Y IS A K1*K3 MATRIX
C ------------------------------------------------------------
   DO 30 i = 1, K1
     DO 20 k = 1, K3
       sum = 0.
       DO 10 j = 1, K2
         sum = sum + X(i,j)*Y(j,k)
   10    CONTINUE
       PROD(i,k) = sum
   20  CONTINUE
   30 CONTINUE
   RETURN
   END
C
C=================================================== STUTP    =======     688**
   FUNCTION STUTP(X,N)                             689**
C                                              690**
C STUDENT T PROBABILITY                         691**
C STUTP = PROB( STUDENT T WITH N DEG FR .LT. X )         692**
C                                              693**
C NOTE - PROB(ABS(T).GT.X) = 2.*STUTP(-X,N) (FOR X .GT. 0.)      694**
C                                              695**
C SUBPGM USED - GAUSCF                          696**
C                                              697**
C REF - G.W. HILL, ACM ALGOR 395, OCTOBER 1970.             698**
C                                              699**
C  USGS - WK 12/79.                             700**
C                                              701**
C                                              702**
   DATA RHPI / 0.63661977 /                        703**
C                                              704**
   STUTP = .5                                  705**
   IF(N.LT.1) RETURN                             706**
C                                              707**
   NN = N                                     708**
   Z = 1.                                      709**
   T = X**2                                    710**
   Y = T/NN                                    711**
   B = 1.0 + Y                                  712**
C                                              713**
   IF(NN.GE.20 .AND. T.LT.NN .OR. NN.GT.200) GO TO 200       714**
C      ( OR IF NN NON-INTEGER)                      715**
C                                              716**
   IF(NN.LT.20 .AND. T.LT.4.) GO TO 100              717**
C                                              718**
C -- TAIL SERIES FOR LARGE T                       719**
   A = SQRT(B)                                 720**
   Y = A*NN                                    721**
   J = 0                                      722**
   30 J = J + 2                                  723**
   IF(A.EQ.Z) GO TO 40                           724**
   Z = A                                      725**
   Y = Y*(J-1)/(B*J)                             726**
   A = A + Y/(NN+J)                             727**
   GO TO 30                                    728**
```

```
   40 CONTINUE                                             729**
      NN = NN + 2                                          730**
      Z = 0.                                               731**
      Y = 0.                                               732**
      A = -A                                               733**
      GO TO 110                                            734**
C                                               735**
C -- NESTED SUMMATION OF COSINE SERIES                     736**
  100 Y = SQRT(Y)                                 737**
      A = Y                                       738**
      IF(NN.EQ. 1) A = 0.                             739**
  110 NN = NN - 2                                   740**
      IF(NN.LE.1) GO TO 120                            741**
      A = (NN-1)/(B*NN)*A + Y                           742**
      GO TO 110                                   743**
  120 IF(NN.EQ.0) A = A/SQRT(B)                            744**
      IF(NN.NE.0) A = (ATAN(Y)+A/B)*RHPI                     745**
      STUTP = 0.5*(Z-A)                               746**
      IF(X.GT.0.) STUTP = 1.-STUTP                        747**
      RETURN                                     748**
C                                           749**
C -- ASYMPTOTIC SERIES FOR LARGE OR NONINTEGER N             750**
  200 IF(Y.GT.1E-6) Y = ALOG(B)                       751**
      A = NN - 0.5                                752**
      B = 48.*A**2                                753**
      Y = A*Y                                 754**
      Y = ((((((-0.4*Y-3.3)*Y-24.)*Y-85.5)/                755**
     $   (0.8*Y**2+100.+B)+Y+3.)/B+1.)*SQRT(Y)              756**
      STUTP = GAUSCF(-Y)                             757**
      IF(X.GT.0.) STUTP = 1.-STUTP                        758**
      RETURN                                     759**
C                                           760**
      END                                     761**
C=================================================             762**
c
C=================================================== GAUSEX    ======   1213**
      FUNCTION GAUSEX(EXPROB)                         1214**
C                                           1215**
C GAUSSIAN PROBABILITY FUNCTIONS  W.KIRBY JUNE 71              1216**
C   GAUSEX=VALUE EXCEEDED WITH PROB EXPROB                   1217**
C   GAUSAB=VALUE (NOT EXCEEDED) WITH PROBCUMPROB             1218**
C   GAUSCF=CUMULATIVE PROBABILITY FUNCTION                  1219**
C   GAUSDY=DENSITY FUNCTION                          1220**
C SUBPGMS USED -- NONE                              1221**
C                                           1222**
C GAUSCF MODIFIED 740906 WK -- REPLACED ERF FCN REF BY RATIONAL APPRX N    1223**
C  ALSO REMOVED DOUBLE PRECISION FROM GAUSEX AND GAUSAB.           1224**
C 76-05-04 WK -- TRAP UNDERFLOWS IN EXP IN GUASCF AND DY.         1225**
C                                           1226**
C                                           1227**
      DATA XLIM / 18.3 /                             1228**
      DATAC0,C1,C2/2.51551700, .8028530000, .0103280000/       1229**
      DATAD1,D2,D3/1.432788000, .1892690000, .0013080000/       1230**
C                                           1231**
      P=EXPROB                                   1232**
    2 IF(P.LT.1.0)GOTO10                             1233**
      GAUSEX=-10.                                 1234**
      RETURN                                     1235**
   10 IF(P.GT.0.)GOTO20                             1236**
      GAUSEX=+10.                                 1237**
      RETURN                                     1238**
   20 PR=P                                     1239**
      IF(P.GT..5)PR=1.00-PR                           1240**
      T= SQRT(-2.00*ALOG(PR))                          1241**
      GAUSEX=T-(C0+T*(C1+T*C2))/(1.D0+T*(D1+T*(D2+T*D3)))           1242**
      IF(P.GT..5)GAUSEX=-GAUSEX                         1243**
      RETURN                                     1244**
```

```
C                                          1245**
   ENTRYGAUSAB(CUMPRB)                        1246**
   GAUSAB = 0.                             1247**
   P=1.-CUMPRB                             1248**
   GOTO2                                   1249**
C                                          1250**
   ENTRY GAUSCF(XX)                          1251**
   AX=ABS(XX)                              1252**
   GAUSCF=1.                               1253**
   IF(AX.GT.XLIM)GOTO101                     1254**
   T=1.0/(1.0+.2316419*AX)                   1255**
   D=0.3989423*EXP(-XX*XX*.5)                 1256**
   GAUSCF=1.-D*T*((((1.330274*T - 1.821256)*T + 1.781478)*T -      1257**
   1 0.3565638)*T + 0.3193815)                 1258**
 101 IF(XX.LT.0)GAUSCF=1.-GAUSCF                  1259**
   RETURN                                  1260**
C                                          1261**
   ENTRY GAUSDY (XX)                          1262**
   GAUSDY=0.                               1263**
   IF(ABS(XX).GT.XLIM) RETURN                   1264**
   GAUSDY=.3989423*EXP(-.500*XX*XX)             1265**
   RETURN                                  1266**
   END                                     1267**
C==================================================            1268**
```

www.ingramcontent.com/pod-product-compliance
Lightning Source LLC
Chambersburg PA
CBHW081408170526
45166CB00010B/3250